THE COMMON CORE

Clarifying Expectations
for Teachers & Students

MATH

Grade 6

**Created and Presented by
Align, Assess, Achieve**

Mc Graw Hill Education

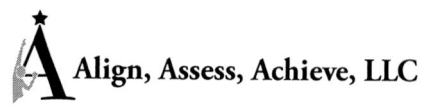

Align, Assess, Achieve, LLC

Align, Assess, Achieve; *The Common Core: Clarifying Expectations for Teachers & Students.* Grade 6

STEM McGraw-Hill is committed to providing instructional materials in Science, Technology, Engineering, and Mathematics (STEM) that give all students a solid foundation, one that prepares them for college and careers in the 21st century.

Send all inquiries to:
McGraw-Hill Education
STEM Learning Solutions Center
8787 Orion Place
Columbus, OH 43240

ISBN: 978-007-662898-8
MHID: 0-07-662898-1

Printed in the United States of America.

1 2 3 4 5 6 7 8 9 GLO 16 15 14 13 12 11

STEM

Our mission is to provide educational resources that enable students to become the problem solvers of the 21st century and inspire them to explore careers within Science, Technology, Engineering, and Mathematics (STEM) related fields.

The McGraw·Hill Companies

Acknowledgements

This book integrates the Common Core State Standards – a framework for educating students to be competitive at an international level – with well-researched instructional planning strategies for achieving the goals of the CCSS. Our work is rooted in the thinking of brilliant educators, such as Grant Wiggins, Jay McTighe, and Rick Stiggins, and enriched by our work with a great number of inspiring teachers, administrators, and parents. We hope this book provides a meaningful contribution to the ongoing conversation around educating lifelong, passionate learners.

We would like to thank many talented contributors who helped create *The Common Core: Clarifying Expectations for Teachers and Students*. Our author Laura Hance, for her intelligence, persistence, and love of teaching; Graphic Designer Thomas Davis, for his creative talents and good nature through many trials; Editors, Sandra Baker, Dr. Teresa Dempsey, and Wesley Yuu, for their educational insight and deep understanding of mathematics; Director of book editing and production Josh Steskal, for his feedback, organization, and unwavering patience; Our spouses, Andrew Bainbridge and Tawnya Holman, who believe in our mission and have, through their unconditional support and love, encouraged us to take risks and grow.

Katy Bainbridge
Bob Holman
Co-Founders
Align, Assess, Achieve, LLC

Executive Editors: *Katy Bainbridge, Bob Holman, Sandra Baker, and Wesley Yuu*
Contributing Authors: *Deborah L. Kaiser, Theresa Mariea, Laura Hance, Ali Fleming, Melissa L. McCreary, Charles L. Brads, Teresa Dempsey, Rebecca Watkins-Heinze, Bob Holman, Wesley Yuu*
Editors: *Jason Bates, Charles L. Brads, Marisa Hilvert, Stephanie Archer*
Graphic Design & Layout: *Thomas Davis; thomasanceldesign.com*
Director of Book Editing & Production: *Josh Steskal*

Introduction

Purpose

The Common Core State Standards (CCSS) provide educators across the nation with a shared vision for student achievement. They also provide a shared challenge: how to interpret the standards and use them in a meaningful way? Clarifying the Common Core was designed to facilitate the transition to the CCSS at the district, building, and classroom level.

Organization

Clarifying the Common Core presents content from two sources: the CCSS and Align, Assess, Achieve. Content from the CCSS is located in the top section of each page and includes the domain, cluster, and grade level standard. The second section of each page contains content created by Align, Assess, Achieve – Enduring Understandings, Essential Questions, Suggested Learning Targets, and Vocabulary. The black bar at the bottom of the page contains the CCSS standard identifier. A sample page can be found in the next section.

Planning for Instruction and Assessment

This book was created to foster meaningful instruction of the CCSS. This requires planning both quality instruction and assessment. Designing and using quality assessments is key to high-quality instruction (Stiggins et al.). Assessment should accurately measure the intended learning and should inform further instruction. This is only possible when teachers and students have a clear vision of the intended learning. When planning instruction it helps to ask two questions, "Where am I taking my students?" and "How will we get there?" The first question refers to the big picture and is addressed with **Enduring Understandings** and **Essential Questions**. The second question points to the instructional process and is addressed by **Learning Targets**.

Where Am I Taking My Students?

When planning, it is useful to think about the larger, lasting instructional concepts as **Enduring Understandings**. Enduring Understandings are rooted in multiple units of instruction throughout the year and are often utilized K-12. These concepts represent the lasting understandings that transcend your content. Enduring Understandings serve as the ultimate goal of a teacher's instructional planning. Although tempting to share with students initially, we do not recommend telling students the Enduring Understanding in advance. Rather, Enduring Understandings are developed through meaningful engagement with an Essential Question.

(continued on next page)

Essential Questions work in concert with Enduring Understandings to ignite student curiosity. These questions help students delve deeper and make connections between the concepts and the content they are learning. Essential Questions are designed with the student in mind and do not have an easy answer; rather, they are used to spark inquiry into the deeper meanings (Wiggins and McTighe). Therefore, we advocate frequent use of Essential Questions with students. It is sometimes helpful to think of the Enduring Understanding as the answer to the Essential Question.

How Will We Get There?

If Enduring Understandings and Essential Questions represent the larger, conceptual ideas, then what guides the learning of specific knowledge, reasoning, and skills? These are achieved by using **Learning Targets**. Learning Targets represent a logical, student friendly progression of teaching and learning. Targets are the scaffolding students climb as they progress towards deeper meaning.

There are four types of learning targets, based on what students are asked to do: knowledge, reasoning/understanding, skill, and product (Stiggins et al.). When selecting Learning Targets, teachers need to ask, "What is the goal of instruction?" After answering this question, select the learning target or targets that align to the instructional goal.

Instructional Goal	*Types of Learning Targets*	*Key Verbs*
Recall basic information and facts	Knowledge (K)	Name, identify, describe
Think and develop an understanding	Reasoning/Understanding (R)	Explain, compare and contrast, predict
Apply knowledge and reasoning	Skill (S)	Use, solve, calculate
Synthesize to create original work	Product (P)	Create, write, present

Adapted from Stiggins et al. *Classroom Assessment for Student Learning.* (Portland: ETS, 2006). Print.

Each book contains two types of Enduring Understandings and Essential Questions. The first type, located on the inside cover, relate to the Mathematical Practices, which apply K-12. The second type are based on the domain, cluster, and standard and are located beneath each standard.

Keep in mind that the Enduring Understandings, Essential Questions, and Learning Targets in this book are suggestions. Modify and combine the content as necessary to meet your instructional needs. Quality instruction consists of clear expectations, ongoing assessment, and effective feedback. Taken together, these promote meaningful instruction that facilitates student mastery of the Common Core State Standards.

References

Stiggins, Rick, Jan Chappuis, Judy Arter, and Steve Chappuis. *Classroom Assessment for Student Learning.* 2nd. Portland, OR: ETS, 2006.

Wiggins, Grant, and Jay McTighe. *Understanding by Design, Expanded 2nd Edition.* 2nd. Alexandria, VA: ASCD, 2005.

Page Organization

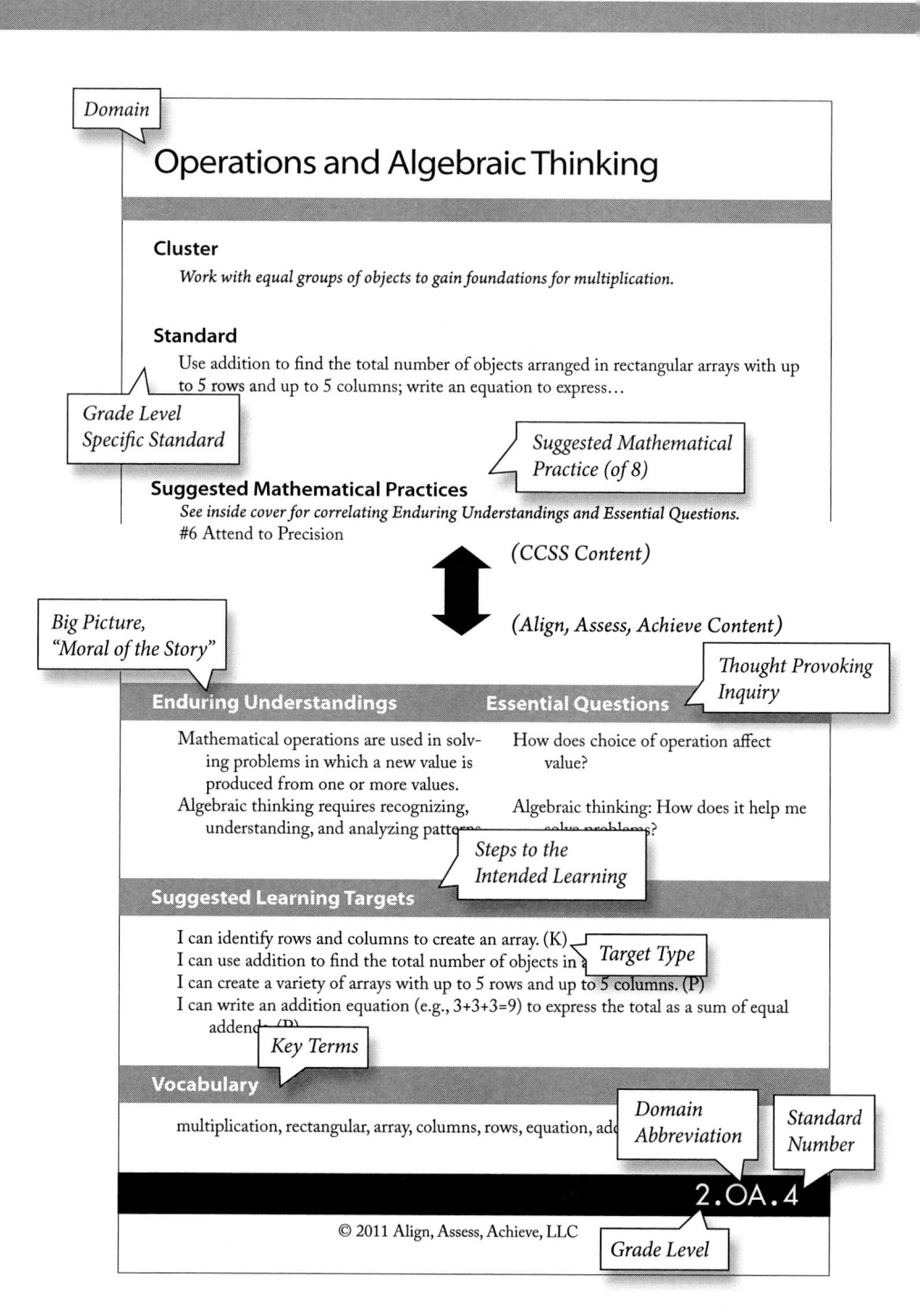

Domain

Operations and Algebraic Thinking

Cluster
Work with equal groups of objects to gain foundations for multiplication.

Standard
Use addition to find the total number of objects arranged in rectangular arrays with up to 5 rows and up to 5 columns; write an equation to express...

Grade Level Specific Standard

Suggested Mathematical Practice (of 8)

Suggested Mathematical Practices
See inside cover for correlating Enduring Understandings and Essential Questions.
#6 Attend to Precision

(CCSS Content)

(Align, Assess, Achieve Content)

Big Picture, "Moral of the Story"

Thought Provoking Inquiry

Enduring Understandings	Essential Questions
Mathematical operations are used in solving problems in which a new value is produced from one or more values.	How does choice of operation affect value?
Algebraic thinking requires recognizing, understanding, and analyzing patterns.	Algebraic thinking: How does it help me solve problems?

Steps to the Intended Learning

Suggested Learning Targets

I can identify rows and columns to create an array. (K)
I can use addition to find the total number of objects in
I can create a variety of arrays with up to 5 rows and up to 5 columns. (P)
I can write an addition equation (e.g., 3+3+3=9) to express the total as a sum of equal addends. (P)

Target Type

Key Terms

Vocabulary

multiplication, rectangular, array, columns, rows, equation, add

Domain Abbreviation

Standard Number

2.OA.4

© 2011 Align, Assess, Achieve, LLC

Grade Level

Mathematical Practices

#1 Making sense of problems and persevere in solving them.

Mathematically proficient students start by explaining to themselves the meaning of a problem and looking for entry points to its solution. They analyze givens, constraints, relationships, and goals. They make conjectures about the form and meaning of the solution and plan a solution pathway rather than simply jumping into a solution attempt. They consider analogous problems, and try special cases and simpler forms of the original problem in order to gain insight into its solution. They monitor and evaluate their progress and change course if necessary. Older students might, depending on the context of the problem, transform algebraic expressions or change the viewing window on their graphing calculator to get the information they need. Mathematically proficient students can explain correspondences between equations, verbal descriptions, tables, and graphs or draw diagrams of important features and relationships, graph data, and search for regularity or trends. Younger students might rely on using concrete objects or pictures to help conceptualize and solve a problem. Mathematically proficient students check their answers to problems using a different method, and they continually ask themselves, "Does this make sense?" They can understand the approaches of others to solving complex problems and identify correspondences between different approaches.

#2 Reason abstractly and quantitatively.

Mathematically proficient students make sense of quantities and their relationships in problem situations. They bring two complementary abilities to bear on problems involving quantitative relationships: the ability to *decontextualize*—to abstract a given situation and represent it symbolically and manipulate the representing symbols as if they have a life of their own, without necessarily attending to their referents—and the ability to *contextualize*, to pause as needed during the manipulation process in order to probe into the referents for the symbols involved. Quantitative reasoning entails habits of creating a coherent representation of the problem at hand; considering the units involved; attending to the meaning of quantities, not just how to compute them; and knowing and flexibly using different properties of operations and objects.

#3. Construct viable arguments and critique the reasoning of others.

Mathematically proficient students understand and use stated assumptions, definitions, and previously established results in constructing arguments. They make conjectures and build a logical progression of statements to explore the truth of their conjectures. They are able to analyze situations by breaking them into cases, and can recognize and use counterexamples. They justify their conclusions, communicate them to others, and respond to the arguments of others. They reason inductively about data, making plausible arguments that take into account the context from which the data arose. Mathematically proficient students are also

(continued on next page)

able to compare the effectiveness of two plausible arguments, distinguish correct logic or reasoning from that which is flawed, and—if there is a flaw in an argument—explain what it is. Elementary students can construct arguments using concrete referents such as objects, drawings, diagrams, and actions. Such arguments can make sense and be correct, even though they are not generalized or made formal until later grades. Later, students learn to determine domains to which an argument applies. Students at all grades can listen or read the arguments of others, decide whether they make sense, and ask useful questions to clarify or improve the arguments.

#4 Model with mathematics.

Mathematically proficient students can apply the mathematics they know to solve problems arising in everyday life, society, and the workplace. In early grades, this might be as simple as writing an addition equation to describe a situation. In middle grades, a student might apply proportional reasoning to plan a school event or analyze a problem in the community. By high school, a student might use geometry to solve a design problem or use a function to describe how one quantity of interest depends on another. Mathematically proficient students who can apply what they know are comfortable making assumptions and approximations to simplify a complicated situation, realizing that these may need revision later. They are able to identify important quantities in a practical situation and map their relationships using such tools as diagrams, two-way tables, graphs, flowcharts and formulas. They can analyze those relationships mathematically to draw conclusions. They routinely interpret their mathematical results in the context of the situation and reflect on whether the results make sense, possibly improving the model if it has not served its purpose.

#5 Use appropriate tools strategically.

Mathematically proficient students consider the available tools when solving a mathematical problem. These tools might include pencil and paper, concrete models, a ruler, a protractor, a calculator, a spreadsheet, a computer algebra system, a statistical package, or dynamic geometry software. Proficient students are sufficiently familiar with tools appropriate for their grade or course to make sound decisions about when each of these tools might be helpful, recognizing both the insight to be gained and their limitations. For example, mathematically proficient high school students analyze graphs of functions and solutions generated using a graphing calculator. They detect possible errors by strategically using estimation and other mathematical knowledge. When making mathematical models, they know that technology can enable them to visualize the results of varying assumptions, explore consequences, and compare predictions with data. Mathematically proficient students at various grade levels are able to identify relevant external mathematical resources, such as digital content located on a website, and use them to pose or solve problems. They are able to use technological tools to explore and deepen their understanding of concepts.

(continued on next page)

#6 Attend to precision.

Mathematically proficient students try to communicate precisely to others. They try to use clear definitions in discussion with others and in their own reasoning. They state the meaning of the symbols they choose, including using the equal sign consistently and appropriately. They are careful about specifying units of measure, and labeling axes to clarify the correspondence with quantities in a problem. They calculate accurately and efficiently, express numerical answers with a degree of precision appropriate for the problem context. In the elementary grades, students give carefully formulated explanations to each other. By the time they reach high school they have learned to examine claims and make explicit use of definitions.

#7 Look for and make use of structure.

Mathematically proficient students look closely to discern a pattern or structure. Young students, for example, might notice that three and seven more is the same amount as seven and three more, or they may sort a collection of shapes according to how many sides the shapes have. Later, students will see 7×8 equals the well remembered $7 \times 5 + 7 \times 3$, in preparation for learning about the distributive property. In the expression $x^2 + 9x + 14$, older students can see the 14 as 2×7 and the 9 as $2 + 7$. They recognize the significance of an existing line in a geometric figure and can use the strategy of drawing an auxiliary line for solving problems. They also can step back for an overview and shift perspective. They can see complicated things, such as some algebraic expressions, as single objects or as being composed of several objects. For example, they can see $5 - 3(x - y)^2$ as 5 minus a positive number times a square and use that to realize that its value cannot be more than 5 for any real numbers x and y.

#8 Look for and express regularity in repeated reasoning.

Mathematically proficient students notice if calculations are repeated, and look both for general methods and for shortcuts. Upper elementary students might notice when dividing 25 by 11 that they are repeating the same calculations over over again, and conclude they have a repeating decimal. By paying attention to the calculation of slope as they repeatedly check whether points are on the line through $(1, 2)$ with slope 3, middle school students might abstract the equation $(y - 2)/(x - 1) = 3$. Noticing the regularity in the way terms cancel when expanding $(x - 1)(x + 1)$, $(x - 1)(x^2 + x + 1)$, and $(x - 1)(x^3 + x^2 + x + 1)$ might lead them to the general formula for the sum of a geometric series. As they work to solve a problem, mathematically proficient students maintain oversight of the process, while attending to the details. They continually evaluate the reasonableness of their intermediate results.

CCSS Grade Level Introduction

In Grade 6, instructional time should focus on four critical areas: (1) connecting ratio and rate to whole number multiplication and division and using concepts of ratio and rate to solve problems; (2) completing understanding of division of fractions and extending the notion of number to the system of rational numbers, which includes negative numbers; (3) writing, interpreting, and using expressions and equations; and (4) developing understanding of statistical thinking.

1. Students use reasoning about multiplication and division to solve ratio and rate problems about quantities. By viewing equivalent ratios and rates as deriving from, and extending, pairs of rows (or columns) in the multiplication table, and by analyzing simple drawings that indicate the relative size of quantities, students connect their understanding of multiplication and division with ratios and rates. Thus students expand the scope of problems for which they can use multiplication and division to solve problems, and they connect ratios and fractions. Students solve a wide variety of problems involving ratios and rates.

2. Students use the meaning of fractions, the meanings of multiplication and division, and the relationship between multiplication and division to understand and explain why the procedures for dividing fractions make sense. Students use these operations to solve problems. Students extend their previous understandings of number and the ordering of numbers to the full system of rational numbers, which includes negative rational numbers, and in particular negative integers. They reason about the order and absolute value of rational numbers and about the location of points in all four quadrants of the coordinate plane.

3. Students understand the use of variables in mathematical expressions. They write expressions and equations that correspond to given situations, evaluate expressions, and use expressions and formulas to solve problems. Students understand that expressions in different forms can be equivalent, and they use the properties of operations to rewrite expressions in equivalent forms. Students know that the solutions of an equation are the values of the variables that make the equation true. Students use properties of operations and the idea of maintaining the equality of both sides of an equation to solve simple one-step equations. Students construct and analyze tables, such as tables of quantities that are in equivalent ratios, and they use equations (such as $3x = y$) to describe relationships between quantities.

(continued on next page)

4. Building on and reinforcing their understanding of number, students begin to develop their ability to think statistically. Students recognize that a data distribution may not have a definite center and that different ways to measure center yield different values. The median measures center in the sense that it is roughly the middle value. The mean measures center in the sense that it is the value that each data point would take on if the total of the data values were redistributed equally, and also in the sense that it is a balance point. Students recognize that a measure of variability (interquartile range or mean absolute deviation) can also be useful for summarizing data because two very different sets of data can have the same mean and median yet be distinguished by their variability.

Students learn to describe and summarize numerical data sets, identifying clusters, peaks, gaps, and symmetry, considering the context in which the data were collected. Students in Grade 6 also build on their work with area in elementary school by reasoning about relationships among shapes to determine area, surface area, and volume. They find areas of right triangles, other triangles, and special quadrilaterals by decomposing these shapes, rearranging or removing pieces, and relating the shapes to rectangles. Using these methods, students discuss, develop, and justify formulas for areas of triangles and parallelograms. Students find areas of polygons and surface areas of prisms and pyramids by decomposing them into pieces whose area they can determine. They reason about right rectangular prisms with fractional side lengths to extend formulas for the volume of a right rectangular prism to fractional side lengths. They prepare for work on scale drawings and constructions in Grade 7 by drawing polygons in the coordinate plane.

Ratios and Proportional Relationships

Cluster

Understand ratio concepts and use ratio reasoning to solve problems.

Standard

Understand the concept of a ratio and use ratio language to describe a ratio relationship between two quantities. *For example, "The ratio of wings to beaks in the bird house at the zoo was 2:1, because for every 2 wings there was 1 beak." "For every vote candidate A received, candidate C received nearly three votes."*

Suggested Mathematical Practices

See inside cover for correlating Enduring Understandings and Essential Questions.
#4 Model with mathematics.

Suggested Learning Targets

I can define the term ratio and demonstrate my understanding by giving various examples. (K)
I can write a ratio that describes a relationship between two quantities. (S)
I can explain the relationship that a ratio represents. (R)

Vocabulary

ratio

Ratios and Proportional Relationships

Cluster
Understand ratio concepts and use ratio reasoning to solve problems.

Standard
Understand the concept of a unit rate a/b associated with a ratio $a:b$ with $b \neq 0$, and use rate language in the context of a ratio relationship. *For example, "This recipe has a ratio of 3 cups of flour to 4 cups of sugar, so there is 3/4 cup of flour for each cup of sugar." "We paid $75 for 15 hamburgers, which is a rate of $5 per hamburger."* *

*Expectations for unit rates in this grade are limited to non-complex fractions.

Suggested Mathematical Practices
See inside cover for correlating Enduring Understandings and Essential Questions.
#2 Reason abstractly and quantitatively.

Enduring Understandings

Ratios and proportional relationships are used to express how quantities are related and how quantities change in relation to each other.

Essential Questions

How can ratios and proportional relationships be used to determine unknown quantities?

Suggested Learning Targets

I can define the term "unit rate" and demonstrate my understanding by giving various examples. (K)
I can recognize a ratio written as a unit rate, explain a unit rate, and give an example of a unit rate. (K)
I can convert a given ratio to a unit rate. (S)
I can describe the ratio relationship represented by a unit rate. (R)

Vocabulary

ratio, rate, unit rate

Ratios and Proportional Relationships

Cluster

Understand ratio concepts and use ratio reasoning to solve problems.

Standard

Use ratio and rate reasoning to solve real-world and mathematical problems, e.g., by reasoning about tables of equivalent ratios, tape diagrams, double number line diagrams, or equations.

   a.   Make tables of equivalent ratios relating quantities with whole-number measurements, find missing values in the tables, and plot the pairs of values on the coordinate plane. Use tables to compare ratios.

   b.   Solve unit rate problems including those involving unit pricing and constant speed. *For example, if it took 7 hours to mow 4 lawns, then at that rate, how many lawns could be mowed in 35 hours? At what rate were lawns being mowed?*

   c.   Find a percent of a quantity as a rate per 100 (e.g., 30% of a quantity means 30/100 times the quantity); solve problems involving finding the whole, given a part and the percent.

   d.   Use ratio reasoning to convert measurement units; manipulate and transform units appropriately when multiplying or dividing quantities.

Suggested Mathematical Practices

See inside cover for correlating Enduring Understandings and Essential Questions.

#2 Reason abstractly and quantitatively.

Enduring Understandings

Ratios and proportional relationships are used to express how quantities are related and how quantities change in relation to each other.

Essential Questions

How can ratios and proportional relationships be used to determine unknown quantities?

Suggested Learning Targets

I can solve real-world problems involving proportional reasoning by using various diagrams. (R)

(continued on next page)

Vocabulary

ratio, equivalent ratio, rate, unit rate, percent, coordinate plane

Ratios and Proportional Relationships

Cluster

Understand ratio concepts and use ratio reasoning to solve problems.

Standard

Use ratio and rate reasoning to solve real-world and mathematical problems, e.g., by reasoning about tables of equivalent ratios, tape diagrams, double number line diagrams, or equations.

a.   Make tables of equivalent ratios relating quantities with whole-number measurements, find missing values in the tables, and plot the pairs of values on the coordinate plane. Use tables to compare ratios.

b.   Solve unit rate problems including those involving unit pricing and constant speed. *For example, if it took 7 hours to mow 4 lawns, then at that rate, how many lawns could be mowed in 35 hours? At what rate were lawns being mowed?*

c.   Find a percent of a quantity as a rate per 100 (e.g., 30% of a quantity means 30/100 times the quantity); solve problems involving finding the whole, given a part and the percent.

d.   Use ratio reasoning to convert measurement units; manipulate and transform units appropriately when multiplying or dividing quantities.

Suggested Mathematical Practices

See inside cover for correlating Enduring Understandings and Essential Questions.

#2 Reason abstractly and quantitatively.

Suggested Learning Targets

(continued from previous page)

I can create a table of equivalent ratios. (S)
I can use the proportional relationship to find missing values in a table of equivalent ratios. (R)
I can compare ratios presented in various tables. (R)
I can plot corresponding values from an equivalent ratio table on a coordinate grid. (S)
I can use proportional reasoning to solve unit rate problems. (R)
I can use visual representations (e.g., strip diagrams, percent bars, one-hundred grids) to model percents. (R)
I can write a percent as a rate per one-hundred. (K)
I can use proportional reasoning to find the percent of a given number. (R)
I can use proportional reasoning to find the whole when given both the part and the percent. (R)
I can use a ratio as a conversion factor when working with measurements of different units. (S)

6 . RP . 3 *(cont.)*

The Number System

Cluster

Apply and extend previous understandings of multiplication and division to divide fractions by fractions.

Standard

Interpret and compute quotients of fractions, and solve word problems involving division of fractions by fractions, e.g., by using visual fraction models and equations to represent the problem. *For example, create a story context for (2/3) ÷ (3/4) and use a visual fraction model to show the quotient; use the relationship between multiplication and division to explain that (2/3) ÷ (3/4) = 8/9 because 3/4 of 8/9 is 2/3. (In general, (a/b) ÷ (c/d) = ad/bc.) How much chocolate will each person get if 3 people share 1/2 lb of chocolate equally? How many 3/4-cup servings are in 2/3 of a cup of yogurt? How wide is a rectangular strip of land with length 3/4 mi and area 1/2 square mi? Compute fluently with multi-digit numbers and find common factors and multiples.*

Suggested Mathematical Practices

See inside cover for correlating Enduring Understandings and Essential Questions.

#4 Model with mathematics.

Enduring Understandings

Rational numbers can be represented in multiple ways and are useful when examining situations involving numbers that are not whole.

Essential Questions

In what ways can rational numbers be useful?

Suggested Learning Targets

I can use a visual model to represent the division of a fraction by a fraction. (S)
I can divide fractions by fractions using an algorithm or mathematical reasoning. (S)
I can justify the quotient of a division problem by relating it to a multiplication problem. (R)
I can use mathematical reasoning to justify the standard algorithm for fraction division. (R)
I can solve real world problems involving the division of fractions and interpret the quotient in the context of the problem. (S)
I can create story contexts for problems involving the division of a fraction by a fraction. (P)

Vocabulary

quotient

The Number System

Cluster
Compute fluently with multi-digit numbers and find common factors and multiples.

Standard
Fluently divide multi-digit numbers using the standard algorithm.

Suggested Mathematical Practices
See inside cover for correlating Enduring Understandings and Essential Questions.
#6 Attend to precision.

Enduring Understandings

Rational numbers can be represented in multiple ways and are useful when examining situations involving numbers that are not whole.

Essential Questions

In what ways can rational numbers be useful?

Suggested Learning Targets

I can use the standard algorithm to fluently divide multi-digit numbers. (S)

Vocabulary

(No applicable vocabulary)

6. NS. 2

The Number System

Cluster

Compute fluently with multi-digit numbers and find common factors and multiples.

Standard

Fluently add, subtract, multiply, and divide multi-digit decimals using the standard algorithm for each operation.

Suggested Mathematical Practices

See inside cover for correlating Enduring Understandings and Essential Questions.

#6 Attend to precision.

Enduring Understandings	Essential Questions
Rational numbers can be represented in multiple ways and are useful when examining situations involving numbers that are not whole.	In what ways can rational numbers be useful?

Suggested Learning Targets

I can fluently add and subtract multi-digit decimals using the standard algorithm. (S)
I can fluently multiply multi-digit decimals using the standard algorithm. (S)
I can fluently divide multi-digit decimals using the standard algorithm. (S)

Vocabulary

(No applicable vocabulary)

6.NS.3

The Number System

Cluster

Compute fluently with multi-digit numbers and find common factors and multiples.

Standard

Find the greatest common factor of two whole numbers less than or equal to 100 and the least common multiple of two whole numbers less than or equal to 12. Use the distributive property to express a sum of two whole numbers 1–100 with a common factor as a multiple of a sum of two whole numbers with no common factor. *For example, express 36 + 8 as 4 (9 + 2). Apply and extend previous understandings of numbers to the system of rational numbers.*

Suggested Mathematical Practices

See inside cover for correlating Enduring Understandings and Essential Questions.
#7 Look for and make use of structure.

Enduring Understandings	Essential Questions
Rational numbers can be represented in multiple ways and are useful when examining situations involving numbers that are not whole.	In what ways can rational numbers be useful?

Suggested Learning Targets

I can find all factors of any given number, less than or equal to 100. (S)
I can find the greatest common factor of any two numbers, less than or equal to 100. (S)
I can create a list of multiples for any number less than or equal to 12. (S)
I can find the least common multiple of any two numbers, less than or equal to 12. (S)
I can use the distributive property to rewrite a simple addition problem when the addends have a common factor. (S)

Vocabulary

factor, multiple, greatest common factor, least common multiple, distributive property

6. NS.4

The Number System

Cluster

Apply and extend previous understandings of numbers to the system of rational numbers.

Standard

Understand that positive and negative numbers are used together to describe quantities having opposite directions or values (e.g., temperature above/below zero, elevation above/below sea level, credits/debits, positive/negative electric charge); use positive and negative numbers to represent quantities in real-world contexts, explaining the meaning of 0 in each situation.

Suggested Mathematical Practices

See inside cover for correlating Enduring Understandings and Essential Questions.
#2 Reason abstractly and quantitatively.

Enduring Understandings

Rational numbers can be represented in multiple ways and are useful when examining situations involving numbers that are not whole.

Essential Questions

In what ways can rational numbers be useful?

Suggested Learning Targets

I can describe and give examples of how positive or negative numbers are used to describe quantities having opposite directions or opposite values. (R)

I can recognize that positive and negative signs represent opposite values and/or directions. (K)

I can explain that the number zero is the point at which direction or value will change. (K)

I can use positive and negative numbers along with zero to represent real world situations. (S)

Vocabulary

positive, negative, opposite

6.NS.5

The Number System

Cluster

Apply and extend previous understandings of numbers to the system of rational numbers.

Standard

Understand a rational number as a point on the number line. Extend number line diagrams and coordinate axes familiar from previous grades to represent points on the line and in the plane with negative number coordinates.

a. Recognize opposite signs of numbers as indicating locations on opposite sides of 0 on the number line; recognize that the opposite of the opposite of a number is the number itself, e.g., $-(-3) = 3$, and that 0 is its own opposite.
b. Understand signs of numbers in ordered pairs as indicating locations in quadrants of the coordinate plane; recognize that when two ordered pairs differ only by signs, the locations of the points are related by reflections across one or both axes.
c. Find and position integers and other rational numbers on a horizontal or vertical number line diagram; find and position pairs of integers and other rational numbers on a coordinate plane.

Suggested Mathematical Practices

See inside cover for correlating Enduring Understandings and Essential Questions.

#7 Look for and make use of structure.

Enduring Understandings

Rational numbers can be represented in multiple ways and are useful when examining situations involving numbers that are not whole.

Essential Questions

In what ways can rational numbers be useful?

Suggested Learning Targets

I can show and explain why every rational number can be represented by a point on a number line. (R)

(continued on next page)

Vocabulary

rational number, integer, opposite, coordinate plane, ordered pair, quadrant, reflection

6. NS.6

The Number System

Cluster

Apply and extend previous understandings of numbers to the system of rational numbers.

Standard

Understand a rational number as a point on the number line. Extend number line diagrams and coordinate axes familiar from previous grades to represent points on the line and in the plane with negative number coordinates.

a. Recognize opposite signs of numbers as indicating locations on opposite sides of 0 on the number line; recognize that the opposite of the opposite of a number is the number itself, e.g., $-(-3) = 3$, and that 0 is its own opposite.

b. Understand signs of numbers in ordered pairs as indicating locations in quadrants of the coordinate plane; recognize that when two ordered pairs differ only by signs, the locations of the points are related by reflections across one or both axes.

c. Find and position integers and other rational numbers on a horizontal or vertical number line diagram; find and position pairs of integers and other rational numbers on a coordinate plane.

Suggested Mathematical Practices

See inside cover for correlating Enduring Understandings and Essential Questions.

#7 Look for and make use of structure.

Suggested Learning Targets

(continued from previous page)

I can plot a number and its opposite on a number line and recognize that they are equidistant from zero. (K)

I can find the opposite of any given number including zero. (K)

I can use the signs of the coordinates to determine the location of an ordered pair in the coordinate plane. (K)

I can reason about the location of two ordered pairs that have the same values but different signs. (R)

I can plot a point on a number line or coordinate plane. (S)

I can read a point from a number line or a coordinate plane. (S)

6.NS.6 (cont.)

The Number System

Cluster

Apply and extend previous understandings of numbers to the system of rational numbers.

Standard

Understand ordering and absolute value of rational numbers.

a. Interpret statements of inequality as statements about the relative position of two numbers on a number line diagram. *For example, interpret –3 > –7 as a statement that –3 is located to the right of –7 on a number line oriented from left to right.*

b. Write, interpret, and explain statements of order for rational numbers in real-world contexts. *For example, write –3 °C > –7 °C to express the fact that –3 °C is warmer than –7 °C.*

c. Understand the absolute value of a rational number as its distance from 0 on the number line; interpret absolute value as magnitude for a positive or negative quantity in a real-world situation. *For example, for an account balance of –30 dollars, write |–30| = 30 to describe the size of the debt in dollars.*

d. Distinguish comparisons of absolute value from statements about order. *For example, recognize that an account balance less than –30 dollars represents a debt greater than 30 dollars.*

Suggested Mathematical Practices

See inside cover for correlating Enduring Understandings and Essential Questions.

#2 Reason abstractly and quantitatively.

Enduring Understandings

Rational numbers can be represented in multiple ways and are useful when examining situations involving numbers that are not whole.

Essential Questions

In what ways can rational numbers be useful?

Suggested Learning Targets

I can describe the relative position of two numbers on a number line when given an inequality. (S)

I can interpret a given inequality in terms of a real world situation. (S)

I can define absolute value as it applies to a number line. (K)

I can describe absolute value as the magnitude of the number in a real world situation. (K)

I can compare between using a signed number and using the absolute value of a signed number when referring to real world situations. (R)

Vocabulary

absolute value, magnitude, rational number, positive, negative

6.NS.7

The Number System

Cluster

Apply and extend previous understandings of numbers to the system of rational numbers.

Standard

Solve real-world and mathematical problems by graphing points in all four quadrants of the coordinate plane. Include use of coordinates and absolute value to find distances between points with the same first coordinate or the same second coordinate.

Suggested Mathematical Practices

See inside cover for correlating Enduring Understandings and Essential Questions.

#1 Make sense of problems and persevere in solving them.

Enduring Understandings

Rational numbers can be represented in multiple ways and are useful when examining situations involving numbers that are not whole.

Essential Questions

In what ways can rational numbers be useful?

Suggested Learning Targets

I can graph points in any quadrant of the coordinate plane to solve real-world and mathematical problems. (S)

I can use absolute values to find the distance between two points with the same x-coordinates or the same y-coordinates. (R)

Vocabulary

coordinate plane, quadrant, coordinates, x-coordinate, y-coordinate, absolute value

6.NS.8

Expressions and Equations

Cluster

Apply and extend previous understandings of arithmetic to algebraic expressions.

Standard

Write and evaluate numerical expressions involving whole-number exponents.

Suggested Mathematical Practices

See inside cover for correlating Enduring Understandings and Essential Questions.
#6 Attend to precision.

Enduring Understandings	Essential Questions
Algebraic expressions and equations are used to model real-life problems and represent quantitative relationships, so that the numbers and symbols can be mindfully manipulated to reach a solution or make sense of the quantitative relationships.	How can algebraic expressions and equations be used to model, analyze, and solve mathematical situations?

Suggested Learning Targets

I can explain the meaning of a number raised to a power. (K)
I can write numerical expressions involving whole-number exponents. (S)
I can evaluate numerical expressions involving whole-number exponents. (S)

Vocabulary

base, exponent, evaluate

6.EE.1

Expressions and Equations

Cluster

Apply and extend previous understandings of arithmetic to algebraic expressions.

Standard

Write, read, and evaluate expressions in which letters stand for numbers.

a. Write expressions that record operations with numbers and with letters standing for numbers. *For example, express the calculation "Subtract y from 5" as 5 – y.*

b. Identify parts of an expression using mathematical terms (sum, term, product, factor, quotient, coefficient); view one or more parts of an expression as a single entity. *For example, describe the expression 2 (8 + 7) as a product of two factors; view (8 + 7) as both a single entity and a sum of two terms.*

c. Evaluate expressions at specific values of their variables. Include expressions that arise from formulas used in real-world problems. Perform arithmetic operations, including those involving whole-number exponents, in the conventional order when there are no parentheses to specify a particular order (Order of Operations). *For example, use the formulas $V = s^3$ and $A = 6 s^2$ to find the volume and surface area of a cube with sides of length $s = 1/2$.*

Suggested Mathematical Practices

See inside cover for correlating Enduring Understandings and Essential Questions.

#6 Attend to precision.

Enduring Understandings

Algebraic expressions and equations are used to model real-life problems and represent quantitative relationships, so that the numbers and symbols can be mindfully manipulated to reach a solution or make sense of the quantitative relationships.

Essential Questions

How can algebraic expression and equations be used to model, analyze, and solve mathematical situations?

Suggested Learning Targets

I can translate a relationship given in words into an algebraic expression. (S)

(continued on next page)

Vocabulary

sum, difference, term, product, factor, quotient, coefficient, arithmetic expression, algebraic expression, substitute, evaluate

6.EE.2

Expressions and Equations

Cluster

Apply and extend previous understandings of arithmetic to algebraic expressions.

Standard

Write, read, and evaluate expressions in which letters stand for numbers.

a.   Write expressions that record operations with numbers and with letters standing for numbers. *For example, express the calculation "Subtract y from 5" as 5 – y.*

b.   Identify parts of an expression using mathematical terms (sum, term, product, factor, quotient, coefficient); view one or more parts of an expression as a single entity. *For example, describe the expression 2 (8 + 7) as a product of two factors; view (8 + 7) as both a single entity and a sum of two terms.*

c.   Evaluate expressions at specific values of their variables. Include expressions that arise from formulas used in real-world problems. Perform arithmetic operations, including those involving whole-number exponents, in the conventional order when there are no parentheses to specify a particular order (Order of Operations). *For example, use the formulas $V = s^3$ and $A = 6 s^2$ to find the volume and surface area of a cube with sides of length s = 1/2.*

Suggested Mathematical Practices

See inside cover for correlating Enduring Understandings and Essential Questions.

#6 Attend to precision.

Suggested Learning Targets

(continued from previous page)

I can identify parts of an algebraic expression by using correct mathematical terms. (K)

I can recognize when an expression is representing a sum and/or difference of terms versus a product and/or quotient of terms (e.g., the expression $5(x + 3)$ is representing a product of the terms 5 and $(x + 3)$ while the expression $5x + 3$ is representing a sum of the terms $5x$ and 3). (K)

I can recognize an expression as both a single value and as two or more terms on which an operation is performed. (R)

I can evaluate an algebraic expression for a given value. (S)

I can substitute values in formulas to solve real-world problems. (S)

I can apply the order of operations when evaluating both arithmetic and algebraic expressions. (S)

Expressions and Equations

Cluster

Apply and extend previous understandings of arithmetic to algebraic expressions.

Standard

Apply the properties of operations to generate equivalent expressions. *For example, apply the distributive property to the expression 3 (2 + x) to produce the equivalent expression 6 + 3x; apply the distributive property to the expression 24x + 18y to produce the equivalent expression 6 (4x + 3y); apply properties of operations to y + y + y to produce the equivalent expression 3y.*

Suggested Mathematical Practices

See inside cover for correlating Enduring Understandings and Essential Questions.

#8 Look for and express regularity in repeated reasoning.

Enduring Understandings	Essential Questions
Algebraic expressions and equations are used to model real-life problems and represent quantitative relationships, so that the numbers and symbols can be mindfully manipulated to reach a solution or make sense of the quantitative relationships.	How can algebraic expression and equations be used to model, analyze, and solve mathematical situations?

Suggested Learning Targets

- I can create a visual model to show two expressions are equivalent (e.g., use algebra tiles to model that $3(2 + x) = 6 + 3x$). (R)
- I can apply the properties of operations—especially the distributive property—to generate equivalent expressions. (S)

Vocabulary

equivalent expressions, commutative property, associative property, distributive property

Expressions and Equations

Cluster
Apply and extend previous understandings of arithmetic to algebraic expressions.

Standard
Identify when two expressions are equivalent (i.e., when the two expressions name the same number regardless of which value is substituted into them). *For example, the expressions y + y + y and 3y are equivalent because they name the same number regardless of which number y stands for.*

Suggested Mathematical Practices
See inside cover for correlating Enduring Understandings and Essential Questions.
#8 Look for and express regularity in repeated reasoning.

Enduring Understandings

Algebraic expressions and equations are used to model real-life problems and represent quantitative relationships, so that the numbers and symbols can be mindfully manipulated to reach a solution or make sense of the quantitative relationships.

Essential Questions

How can algebraic expression and equations be used to model, analyze, and solve mathematical situations?

Suggested Learning Targets

I can determine whether two expressions are equivalent by using the same value to evaluate both expressions. (S)
I can use the properties of operations to justify that two expressions are equivalent. (S)

Vocabulary

equivalent expressions

6.EE.4

Expressions and Equations

Cluster
Reason about and solve one-variable equations and inequalities.

Standard
Understand solving an equation or inequality as a process of answering a question: which values from a specified set, if any, make the equation or inequality true? Use substitution to determine whether a given number in a specified set makes an equation or inequality true.

Suggested Mathematical Practices
See inside cover for correlating Enduring Understandings and Essential Questions.
#2 Reason abstractly and quantitatively.

Enduring Understandings

Algebraic expressions and equations are used to model real-life problems and represent quantitative relationships, so that the numbers and symbols can be mindfully manipulated to reach a solution or make sense of the quantitative relationships.

Essential Questions

How can algebraic expression and equations be used to model, analyze, and solve mathematical situations?

Suggested Learning Targets

I can explain that solving an equation or inequality leads to finding the value or values of the variable that will make a true mathematical statement. (R)
I can substitute a given value into an algebraic equation or inequality to determine whether it is part of the solution set. (S)

Vocabulary

equation, inequality, substitute, solve, solution

6.EE.5

Expressions and Equations

Cluster

Reason about and solve one-variable equations and inequalities.

Standard

Use variables to represent numbers and write expressions when solving a real-world or mathematical problem; understand that a variable can represent an unknown number, or, depending on the purpose at hand, any number in a specified set.

Suggested Mathematical Practices

See inside cover for correlating Enduring Understandings and Essential Questions.

#2 Reason abstractly and quantitatively.

Enduring Understandings

Algebraic expressions and equations are used to model real-life problems and represent quantitative relationships, so that the numbers and symbols can be mindfully manipulated to reach a solution or make sense of the quantitative relationships.

Essential Questions

How can algebraic expression and equations be used to model, analyze, and solve mathematical situations?

Suggested Learning Targets

I can use a variable to write an algebraic expression that represents a real-world situation when a specific number is unknown. (S)

I can explain and give examples of how a variable can represent a single unknown number (e.g., $x = 9$, or $5y = 10$) or can represent any number in a specified set (e.g., $m < 8$ or $n + 6 > 10$). (R)

I can use a variable to write an expression that represents a consistent relationship in a particular pattern (e.g., use function tables to write an expression that would represent the output for any input). (S)

Vocabulary

variable, constant, algebraic expression

Expressions and Equations

Cluster

Reason about and solve one-variable equations and inequalities.

Standard

Solve real-world and mathematical problems by writing and solving equations of the form $x + p = q$ and $px = q$ for cases in which p, q and x are all nonnegative rational numbers.

Suggested Mathematical Practices

See inside cover for correlating Enduring Understandings and Essential Questions.

#2 Reason abstractly and quantitatively.

Enduring Understandings

Algebraic expressions and equations are used to model real-life problems and represent quantitative relationships, so that the numbers and symbols can be mindfully manipulated to reach a solution or make sense of the quantitative relationships.

Essential Questions

How can algebraic expression and equations be used to model, analyze, and solve mathematical situations?

Suggested Learning Targets

I can solve equations in the form $x + p = q$ where p and q are given numbers.(S)
I can solve equations in the form $px = q$ where p and q are given numbers. (S)
I can write and solve algebraic equations that represent real world problems. (S)

Vocabulary

algebraic equation, solve

6.EE.7

Expressions and Equations

Cluster

Reason about and solve one-variable equations and inequalities.

Standard

Write an inequality of the form $x > c$ or $x < c$ to represent a constraint or condition in a real-world or mathematical problem. Recognize that inequalities of the form $x > c$ or $x < c$ have infinitely many solutions; represent solutions of such inequalities on number line diagrams.

Suggested Mathematical Practices

See inside cover for correlating Enduring Understandings and Essential Questions.

#2 Reason abstractly and quantitatively.

Enduring Understandings	Essential Questions
Algebraic expressions and equations are used to model real-life problems and represent quantitative relationships, so that the numbers and symbols can be mindfully manipulated to reach a solution or make sense of the quantitative relationships.	How can algebraic expression and equations be used to model, analyze, and solve mathematical situations?

Suggested Learning Targets

I can write a simple inequality to represent the constraints or conditions of numerical values in a real-world or mathematical problem. (S)
I can explain what the solution set of an inequality represents. (R)
I can show the solution set of an inequality by graphing it on a number line. (K)

Vocabulary

inequality

6.EE.8

Expressions and Equations

Cluster

Represent and analyze quantitative relationships between dependent and independent variables.

Standard

Use variables to represent two quantities in a real-world problem that change in relationship to one another; write an equation to express one quantity, thought of as the dependent variable, in terms of the other quantity, thought of as the independent variable. Analyze the relationship between the dependent and independent variables using graphs and tables, and relate these to the equation. *For example, in a problem involving motion at constant speed, list and graph ordered pairs of distances and times, and write the equation $d = 65t$ to represent the relationship between distance and time.*

Suggested Mathematical Practices

See inside cover for correlating Enduring Understandings and Essential Questions.

#1 Make sense of problems and persevere in solving them.

Suggested Learning Targets

I can create a table of two variables that represents a real-world situation in which one quantity will change in relation to the other. (S)

(continued on next page)

Vocabulary

independent variable, dependent variable, coordinate plane

6.EE.9

Expressions and Equations

Cluster

Represent and analyze quantitative relationships between dependent and independent variables.

Standard

Use variables to represent two quantities in a real-world problem that change in relationship to one another; write an equation to express one quantity, thought of as the dependent variable, in terms of the other quantity, thought of as the independent variable. Analyze the relationship between the dependent and independent variables using graphs and tables, and relate these to the equation. *For example, in a problem involving motion at constant speed, list and graph ordered pairs of distances and times, and write the equation $d = 65t$ to represent the relationship between distance and time.*

Suggested Mathematical Practices

See inside cover for correlating Enduring Understandings and Essential Questions.

#1 Make sense of problems and persevere in solving them.

Suggested Learning Targets

(continued from previous page)

I can explain the difference between the independent variable and the dependent variable and give examples of both. (R)

I can determine the independent and dependent variable in a relationship. (S)

I can write an algebraic equation that represents the relationship between the two variables. (S)

I can create a graph by plotting the dependent variable on the x-axis and the independent variable on the y-axis of a coordinate plane. (S)

I can analyze the relationship between the dependent and independent variables by comparing the table, graph, and equation. (R)

6.EE.9 *(cont.)*

Geometry

Cluster
Solve real-world and mathematical problems involving area, surface area, and volume.

Standard
Find the area of right triangles, other triangles, special quadrilaterals, and polygons by composing into rectangles or decomposing into triangles and other shapes; apply these techniques in the context of solving real-world and mathematical problems.

Suggested Mathematical Practices
See inside cover for correlating Enduring Understandings and Essential Questions.
#2 Reason abstractly and quantitatively.

Enduring Understandings

Geometric attributes (such as shapes, lines, angles, figures, and planes) provide descriptive information about an object's properties and position in space and support visualization and problem solving.

Essential Questions

How does geometry better describe objects?

Suggested Learning Targets

I can show how to find the area of a parallelogram by decomposing it and recomposing the parts to form a rectangle. (S)

I can show how to find the area of a right triangle by composing two of them into a rectangle. (S)

I can show how to find the area of a triangle by composing two of them into a parallelogram or rectangle or by decomposing the triangle and recomposing its parts to form a parallelogram or rectangle. (S)

I can show how to find the area of a trapezoid by composing two of them into a rectangle or parallelogram or decomposing the trapezoid into a rectangle and one or more triangles. (S)

I can show how to find the area of other polygons by decomposing them into simpler shapes such as triangles, rectangles, and parallelograms and combining the areas of those simple shapes (S).

I can explain the relationship between the formulas for the area of rectangles, parallelograms, triangles, and trapezoids. (R)

I can solve real-world problems that involve finding the area of polygons. (S)

Vocabulary

polygon, triangle, right triangle, quadrilateral, parallelogram, trapezoid, area, square unit

6.G.1

Geometry

Cluster
Solve real-world and mathematical problems involving area, surface area, and volume.

Standard
Find the volume of a right rectangular prism with fractional edge lengths by packing it with unit cubes of the appropriate unit fraction edge lengths, and show that the volume is the same as would be found by multiplying the edge lengths of the prism. Apply the formulas $V = l\,w\,h$ and $V = b\,h$ to find volumes of right rectangular prisms with fractional edge lengths in the context of solving real-world and mathematical problems.

Suggested Mathematical Practices
See inside cover for correlating Enduring Understandings and Essential Questions.
#2 Reason abstractly and quantitatively.

Enduring Understandings
Geometric attributes (such as shapes, lines, angles, figures, and planes) provide descriptive information about an object's properties and position in space and support visualization and problem solving.

Essential Questions
How does geometry better describe objects?

Suggested Learning Targets
I can find the volume of a right rectangular prism by reasoning about the number of unit cubes it takes to cover the first layer of the prism and the number of layers needed to fill the entire prism. (S)
I can generalize finding the volume of a right rectangular prism to the equation $V=lwh$ or $V=Bh$. (R)
I can solve real-world problems that involve finding the volume of right rectangular prisms. (S)

Vocabulary
right rectangular prism, base, height, area, volume, cubic unit

6.G.2

Geometry

Cluster

Solve real-world and mathematical problems involving area, surface area, and volume.

Standard

Draw polygons in the coordinate plane given coordinates for the vertices; use coordinates to find the length of a side joining points with the same first coordinate or the same second coordinate. Apply these techniques in the context of solving real-world and mathematical problems.

Suggested Mathematical Practices

See inside cover for correlating Enduring Understandings and Essential Questions.
#5 Use appropriate tools strategically.

Enduring Understandings	Essential Questions
Geometric attributes (such as shapes, lines, angles, figures, and planes) provide descriptive information about an object's properties and position in space and support visualization and problem solving.	How does geometry better describe objects?

Suggested Learning Targets

I can plot vertices in the coordinate plane to draw specific polygons. (S)
I can use the coordinates of the vertices of a polygon to find the length of a specific side. (S)
I can plot points, draw figures, and find lengths on the coordinate plane to solve real-world problems. (S)

Vocabulary

vertex/vertices, coordinate, polygon

6.G.3

Geometry

Cluster

Solve real-world and mathematical problems involving area, surface area, and volume.

Standard

Represent three-dimensional figures using nets made up of rectangles and triangles, and use the nets to find the surface area of these figures. Apply these techniques in the context of solving real-world and mathematical problems.

Suggested Mathematical Practices

See inside cover for correlating Enduring Understandings and Essential Questions.

#4 Model with mathematics.

Enduring Understandings	Essential Questions
Geometric attributes (such as shapes, lines, angles, figures, and planes) provide descriptive information about an object's properties and position in space and support visualization and problem solving.	How does geometry better describe objects?

Suggested Learning Targets

I can match a net to the correct right rectangular prism, right triangular prism, right square pyramid, or right tetrahedron. (S)

I can draw a net for a given rectangular prism, right triangular prism, right square pyramid, or right tetrahedron. (S)

I can use a net to find the surface area of a given rectangular prism, right triangular prism, right pyramid, or right tetrahedron. (S)

I can solve real-world problems that involve finding the surface area of a rectangular prism, right triangular prism, right square pyramid, or right tetrahedron. (S)

Vocabulary

right rectangular prism, right triangular prism, right square pyramid, right tetrahedron, net, surface area

6.G.4

Statistics and Probability

Cluster
Develop understanding of statistical variability.

Standard
Recognize a statistical question as one that anticipates variability in the data related to the question and accounts for it in the answers. *For example, "How old am I?" is not a statistical question, but "How old are the students in my school?" is a statistical question because one anticipates variability in students' ages.*

Suggested Mathematical Practices
See inside cover for correlating Enduring Understandings and Essential Questions.
#6 Attend to precision.

Enduring Understandings

The rules of probability can lead to more valid and reliable predictions about the likelihood of an event occurring.

Essential Questions

How is probability used to make informed decisions about uncertain events?

Suggested Learning Targets

I can explain what makes a good statistical question. (R)
I can develop a question that can be used to collect statistical information. (S)

Vocabulary

variability

Statistics and Probability

Cluster

Develop understanding of statistical variability.

Standard

Understand that a set of data collected to answer a statistical question has a distribution which can be described by its center, spread, and overall shape.

Suggested Mathematical Practices

See inside cover for correlating Enduring Understandings and Essential Questions.

#4 Model with mathematics.

Enduring Understandings	Essential Questions
The rules of probability can lead to more valid and reliable predictions about the likelihood of an event occurring.	How is probability used to make informed decisions about uncertain events?

Suggested Learning Targets

I can explain that there are three ways that the distribution of a set of data can be described: by its center, spread, and overall shape. (K)

I can describe the center of a set of statistical data in terms of the mean, median, and the mode. (K)

I can describe the spread of a set of statistical data in terms of extremes, clusters, gaps, and outliers. (K)

I can describe the overall shape of the set of data in terms of its symmetry or skewness. (K)

Vocabulary

distribution, center, spread, shape of data

Statistics and Probability

Cluster
Develop understanding of statistical variability.

Standard
Recognize that a measure of center for a numerical data set summarizes all of its values with a single number, while a measure of variation describes how its values vary with a single number.

Suggested Mathematical Practices
See inside cover for correlating Enduring Understandings and Essential Questions.
#6 Attend to precision.

Enduring Understandings

The rules of probability can lead to more valid and reliable predictions about the likelihood of an event occurring.

Essential Questions

How is probability used to make informed decisions about uncertain events?

Suggested Learning Targets

I can define a measure of center as a single value that summarizes a data set. (K)
I can find measures of center by calculating the mean, median, and mode of a set of numerical data.
I can define a measure of variation as the range of the data, relative to the measures of center. (K)
I can find measures of variation by calculating the interquartile range or the mean absolute deviation of a set of numerical data. (S)

Vocabulary

measure of center, mean, median (Q_2), mode, measure of variation, range, interquartile range, extremes, lower quartile (Q_1), upper quartile(Q_3), outlier, mean absolute deviation

6.SP.3

Statistics and Probability

Cluster

Summarize and describe distributions.

Standard

Display numerical data in plots on a number line, including dot plots, histograms, and box plots.

Suggested Mathematical Practices

See inside cover for correlating Enduring Understandings and Essential Questions.

#4 Model with mathematics.

Suggested Learning Targets

I can organize and display data as a line plot or dot plot. (S)
I can organize and display data in a histogram. (S)
I can organize and display data in a box plot. (S)
I can determine the upper and lower extremes, median, and upper and lower quartiles of a set of data and use this information to display the data in a box plot. (S)
I can identify the similarities and differences of representing the same data in a line plot, a histogram, or a box plot. (R)
I can decide and explain which type of plot (dot plot, line plot, histogram, or box plot) is the best way to display my data depending on what I want to communicate about the data. (R)

Vocabulary

line plot, dot plot, histogram, median (Q_2), lower extreme, lower quartile (Q_1), upper quartile (Q_3), upper extreme, box plot, outlier

Statistics and Probability

Cluster

Summarize and describe distributions.

Standard

Summarize numerical data sets in relation to their context, such as by:
a. Reporting the number of observations.
b. Describing the nature of the attribute under investigation, including how it was measured and its units of measurement.
c. Giving quantitative measures of center (median and/or mean) and variability (interquartile range and/or mean absolute deviation), as well as describing any overall pattern and any striking deviations from the overall pattern with reference to the context in which the data were gathered.
d. Relating the choice of measures of center and variability to the shape of the data distribution and the context in which the data were gathered.

Suggested Mathematical Practices

See inside cover for correlating Enduring Understandings and Essential Questions.
#3 Construct viable arguments and critique the reasoning of others.

Enduring Understandings

The rules of probability can lead to more valid and reliable predictions about the likelihood of an event occurring.

Essential Questions

How is probability used to make informed decisions about uncertain events?

Suggested Learning Targets

I can write a data collection summary that includes the number of observations, what is being investigated, how it is measured, and the units of measurement. (P)
I can determine the measures of center and measures of variability of the collected data. (S)

(continued on next page)

Vocabulary

measure of center, mean, median, mode, measure of variability, range, interquartile range, mean absolute deviation

6.SP.5

Statistics and Probability

Cluster

Summarize and describe distributions.

Standard

Summarize numerical data sets in relation to their context, such as by:

a. Reporting the number of observations.

b. Describing the nature of the attribute under investigation, including how it was measured and its units of measurement.

c. Giving quantitative measures of center (median and/or mean) and variability (interquartile range and/or mean absolute deviation), as well as describing any overall pattern and any striking deviations from the overall pattern with reference to the context in which the data were gathered.

d. Relating the choice of measures of center and variability to the shape of the data distribution and the context in which the data were gathered.

Suggested Mathematical Practices

See inside cover for correlating Enduring Understandings and Essential Questions.

#3 Construct viable arguments and critique the reasoning of others.

Suggested Learning Targets

(continued from previous page)

I can justify the use of a particular measure of center or measure of variability based on the shape of the data. (R)

I can use a measure of center and a measure of variation to draw inferences about the shape of the data distribution. (R)

I can describe overall patterns in the data and how they relate to the context of the problem. (R)

I can describe any deviations from the overall pattern and how they relate to the context of the problem. (R)